Created by Gianluca Lenisa

In the infinite vastness of space, our Solar System shines like a cosmic masterpiece, a harmonious dance of planets, moons, asteroids, and comets around a mother star: the Sun. Welcome to 'Discovering the Solar System', an epic exploration that will lead us through the mysteries and wonders of our celestial neighborhood, in a journey that embraces the science, beauty, and marvel of the universe.

Throughout the centuries, humans have gazed at the night sky, questioning the stars and planets, seeking to unveil the secrets of the cosmos. In this book, we will embark on a journey that will take us from the fiery solar prairies to the outermost boundaries of the outer planets, exploring every corner and curiosity of our Solar System.

Guided by the most renowned astrophysicists and supported by breathtaking images captured by space probes, 'Discovering the Solar System' invites us to peer beyond the horizon, beyond the confines of our current understanding. We will begin our journey from the very heart of the Solar System, where the Sun reigns unchallenged as the pivot of life and the energy that permeates our world and beyond.

Through detailed scientific descriptions and engaging narrative, we will explore the unique features of each planet, from the arid plains of Mercury to the deep valleys of Mars, from the turbulent storms of Jupiter to the icy expanses of Uranus and Neptune. Each celestial body tells a story of formation and evolution, a chapter in the cosmic saga unfolding before our eyes.

But our journey does not stop at the boundaries of the planets. We will also delve into the enigmatic world of asteroids and comets, cosmic travelers that traverse the night sky with their ephemeral yet potentially catastrophic beauty. And we will venture into the icy depths of dwarf planets and trans-Neptunian bodies, silent witnesses to the ancient gravitational dance that has shaped our Solar System.

This book is more than just a collection of astronomical facts; it is an invitation to explore the infinite beauty and complexity of our universe, to be captivated by the majesty of distant galaxies, and to recognize our connection to the stars. Whether you are an astronomy enthusiast or simply curious about the cosmos that surrounds us, 'Discovering the Solar System' will take you on an extraordinary journey that will forever change your perspective on the wonders of the universe. Prepare to gaze skyward and discover our place in the infinite cosmos. Welcome aboard.

In the vast theater of space, the Solar System stands out as one of the most extraordinary and fascinating works of the known universe. Born about 4.6 billion years ago from a vast cloud of cosmic gas and dust, this planetary system is our closest cosmic neighbor, a world of wonders that continues to fascinate and awe humanity.

Comprising eight major planets, a myriad of natural satellites, asteroids, comets, and other celestial bodies, the Solar System is a dynamic microcosm of unique astronomical phenomena. At its center shines our Sun, a medium-sized star that radiates light and heat to fuel life and energy in its surroundings.

The planets of the Solar System are divided into two main categories: the inner planets, mainly composed of rocks and metals, and the outer planets, characterized by gaseous atmospheres and planetary rings. The first four planets - Mercury, Venus, Earth, and Mars - orbit closer to the Sun, while the four gas giants - Jupiter, Saturn, Uranus, and Neptune - lie beyond the asteroid belt.

Each planet of the Solar System carries with it its unique history of formation and evolution, shaped by the forces of gravity, volcanism, solar wind, and cosmic collisions. From the volcanoes of Mars to the giant storms of Jupiter, from the mysterious structures of Saturn's rings to the turbulent atmospheres of Uranus and Neptune, each world offers a fascinating window into the wonders of planetary nature.

In addition to the planets, the Solar System is populated by a vast array of celestial bodies, including moons, asteroids, and comets, each playing a unique role in the cosmic ecosystem. These bodies, along with the planets, compose a cosmic symphony that continues to inspire explorers, scientists, and dreamers around the world.

THE SUN

The Sun is a G2V-type star, commonly known as a yellow dwarf, located at the center of the solar system. Its mass is approximately 1.989×10^{30} kg, representing about 99.8% of the total mass of the solar system.

The surface temperature of the Sun is about 5,500 degrees Celsius, while the temperature at the core reaches about 15 million degrees Celsius, where nuclear fusion reactions occur, transforming hydrogen into helium through the process of nuclear fusion. This reaction generates enormous amounts of energy in the form of electromagnetic radiation.

The brightness of the Sun, which is the total amount of energy emitted per unit of time, is about 3.8×10^{26} watts. This energy is emitted mainly in the form of visible light but also includes ultraviolet, infrared, and other forms of electromagnetic radiation.

The Sun is mainly composed of hydrogen (about 74% of the mass) and helium (about 24% of the mass), with traces of heavier elements such as oxygen, carbon, neon, and iron. Its internal structure consists of several regions, including the core, the radiative zone, and the convective zone.

The gravity of the Sun is responsible for its ability to keep the planets of the solar system in orbit, exerting a gravitational force that determines their orbital motion. The Sun is an essential source of light and heat for Earth and all other celestial bodies in the solar system, playing a crucial role in sustaining life on Earth and in the planet's climatic and atmospheric processes.

MERCURY

Mercury is the innermost planet of the solar system, located at an average distance from the Sun of about 57.9 million kilometers. It is characterized by a rocky and cratered surface, similar to the Moon, with few surface features such as plains, cliffs, and canyons.

Its atmosphere is extremely thin and mainly composed of rarefied gases such as helium, neon, and hydrogen, with traces of oxygen, sodium, hydrogen, and potassium. Due to its proximity to the Sun, Mercury's surface temperature varies widely, ranging from -173 degrees Celsius during the night to 427 degrees Celsius during the day.

Mercury has a very slow rotation on its axis, with a rotation period of about 59 Earth days, but completes one revolution around the Sun in only 88 Earth days. This means that a day on Mercury (from one sunrise to the next) lasts about 176 Earth days.

The planet lacks moons and has a much weaker magnetic field compared to that of Earth. In the past, it has been subject to intense meteorite bombardment, as indicated by the numerous craters on its surface.

Its orbit is slightly eccentric and inclined relative to the plane of the ecliptic. Mercury has been the subject of studies and space missions to better understand its structure, composition, and origin, thus contributing to our understanding of the formation and evolution of the solar system.

VENUS

Venus, the second planet of the solar system, is often called the "twin of Earth" due to its similar size, but its conditions are extremely different. It is a rocky world with a dense atmosphere rich in carbon dioxide and thick layers of clouds primarily composed of sulfuric acid.

The surface temperature of Venus is extremely high, exceeding 450 degrees Celsius, making it the hottest planet in the solar system despite being the second closest planet to the Sun. This extreme greenhouse effect is mainly caused by the greenhouse effect generated by the gases present in the atmosphere, which trap the heat from the Sun.

Venus's surface is characterized by vast lava plains, highlands, and numerous impact craters. It also exhibits a range of unique geological features, such as cone volcanoes, ridges, and deep canyons. One of its distinctive features is the presence of "terrae" or continental regions, mainly in the Ishtar Terra continent region.

Venus rotates very slowly on its axis, with a rotation period of about 243 Earth days, which means that a Venusian day (from one sunrise to the next) is longer than a Venusian year. However, its period of revolution around the Sun is about 225 Earth days.

Despite its proximity to Earth, Venus is a challenging planet to study due to its dense atmosphere and extreme temperatures. However, space missions such as the Magellan probe and ongoing missions such as the European Space Agency's (ESA) Venus Express have provided valuable information about its geology, atmosphere, and climate.

THE EARTH

Earth is the third planet of the solar system, located in the so-called "habitable zone," where conditions are favorable for the presence of liquid water and, consequently, for life as we know it. It is a rocky world, with a surface characterized by landmasses, oceans, and a variety of geological formations.

The Earth's atmosphere is primarily composed of nitrogen (about 78%) and oxygen (about 21%), with traces of other gases such as argon, carbon dioxide, and water vapor. This atmosphere is crucial for life on Earth as it protects the planet from harmful radiation from space and regulates the climate through the water cycle and the greenhouse effect.

The Earth's surface is characterized by continents, oceans, and various geological formations such as mountains, plains, deserts, and volcanoes. It is the only known celestial body to host life, with a vast diversity of organisms inhabiting both on land and in the oceans.

Earth rotates on its axis, completing one full rotation in about 24 hours, thus determining the succession of days and nights. Additionally, it orbits around the Sun in an elliptical orbit, completing one full orbit in about 365.25 days, defining a terrestrial year.

It is important to note that Earth has undergone and continues to undergo climatic and geological changes over time, due to a combination of natural factors and anthropogenic influences. The study of Earth, its history, and its processes is essential for understanding the history of the solar system and for addressing the global challenges that the planet and humanity face in the present and future.

MARS

Mars, known as "the red planet," is the fourth planet of the solar system and is located after Earth. It is a rocky world with a surface characterized by plains, mountains, canyons, and volcanoes, including the largest known volcano, Olympus Mons, and the deepest canyon, Valles Marineris.

The atmosphere of Mars is thin and mainly composed of CO_2 (carbon dioxide), with traces of nitrogen, argon, and oxygen. However, it is significantly thinner than Earth's atmosphere, making the presence of liquid water on the surface difficult. The extreme conditions on the surface of Mars include average temperatures of about -63 degrees Celsius, with peaks of heat and cold depending on the location and time of year.

Mars exhibits geological features that suggest it may have had a warmer and wetter environment in the past, with rivers, lakes, and perhaps even oceans. This is evidenced by the presence of riverbeds and sediments indicating the action of past liquid water.

The presence of water ice has been confirmed in several polar regions and in the form of permafrost in the soil. These water ice deposits could be an important resource for future human missions to Mars.

Mars has a rotation similar to that of Earth, completing one rotation on itself in just over 24 Earth hours, thus defining the duration of a Martian day. Its orbital period around the Sun is about 687 Earth days.

Mars has been the subject of intense exploration by spacecraft and rovers, such as the Viking, Pathfinder, Spirit, Opportunity, Curiosity, and Perseverance missions, which have greatly contributed to our understanding of the planet, its geology, atmosphere, and potential past and future habitability.

JUPITER

Jupiter is the fifth planet of the solar system, as well as the largest. It is a gas giant composed mainly of hydrogen and helium, with traces of other gases and compounds such as ammonia and methane. Its mass is about 318 times that of Earth and constitutes about 70% of the total mass of the planets in the solar system, excluding the Sun.

Jupiter's most distinctive feature is its Great Red Spot, a persistent storm that has been observed for hundreds of years. It is a vast and tumultuous atmospheric structure that can accommodate several terrestrial planets. Its atmosphere also features dark and light atmospheric bands, caused by differences in temperature and chemical composition.

Jupiter is surrounded by a ring system, although they are much weaker and less visible than Saturn's. It also has a large family of moons, with over 80 confirmed moons and many more still being studied and discovered. Jupiter's four largest moons are Io, Europa, Ganymede, and Callisto, known as "the Galilean moons."

Jupiter possesses an extremely powerful magnetic field, about 14 times stronger than that of Earth. This magnetic field is responsible for generating intense radiation of charged particles that create its famous polar auroras.

The internal structure of Jupiter is composed of a rocky core, likely rich in heavy metals, surrounded by a mantle of liquid metallic hydrogen, and, above it, an outer layer of gaseous molecular hydrogen.

Jupiter has been the subject of numerous space missions, including the Pioneer, Voyager, Galileo, and Juno missions, which have greatly contributed to our understanding of this giant planet and its moon system.

SATURN

Saturn is the sixth planet of the solar system, primarily known for its spectacular rings. It is a gas giant, similar to Jupiter, composed mainly of hydrogen and helium, with traces of other gases and compounds such as ammonia and methane. Its mass is about 95 times that of Earth and constitutes about 30% of the total mass of the planets in the solar system, excluding the Sun.

Saturn's most iconic feature is its rings, primarily composed of ice and rock fragments ranging in size from small grains to large boulders. These rings are divided into several main rings, designated with letters of the alphabet, and are separated by empty spaces called divisions. Saturn's rings are the result of gravitational interaction between the planet, its moons, and space debris.

Saturn also has a large family of moons, with over 80 confirmed moons and many more still being studied and discovered. Its largest moon, Titan, is particularly interesting because it has a dense atmosphere and a surface with lakes and seas of liquid hydrocarbons.

Saturn's atmosphere features atmospheric bands similar to those of Jupiter, although less pronounced. Its most distinctive feature is the hexagon-shaped polar storm, a hexagonal structure that extends around the planet's north pole.

Saturn has a significant magnetic field, although not as strong as Jupiter's. However, it is powerful enough to deflect charged particles from the solar wind and create polar auroras.

Saturn's internal structure is similar to that of Jupiter, with a rocky core surrounded by a layer of liquid metallic hydrogen and then an outer layer of gaseous molecular hydrogen.

Saturn has been explored by several space missions, including the Pioneer, Voyager, Cassini-Huygens missions, and the ongoing Cassini mission. These missions have greatly contributed to our understanding of Saturn, its rings, and its moons.

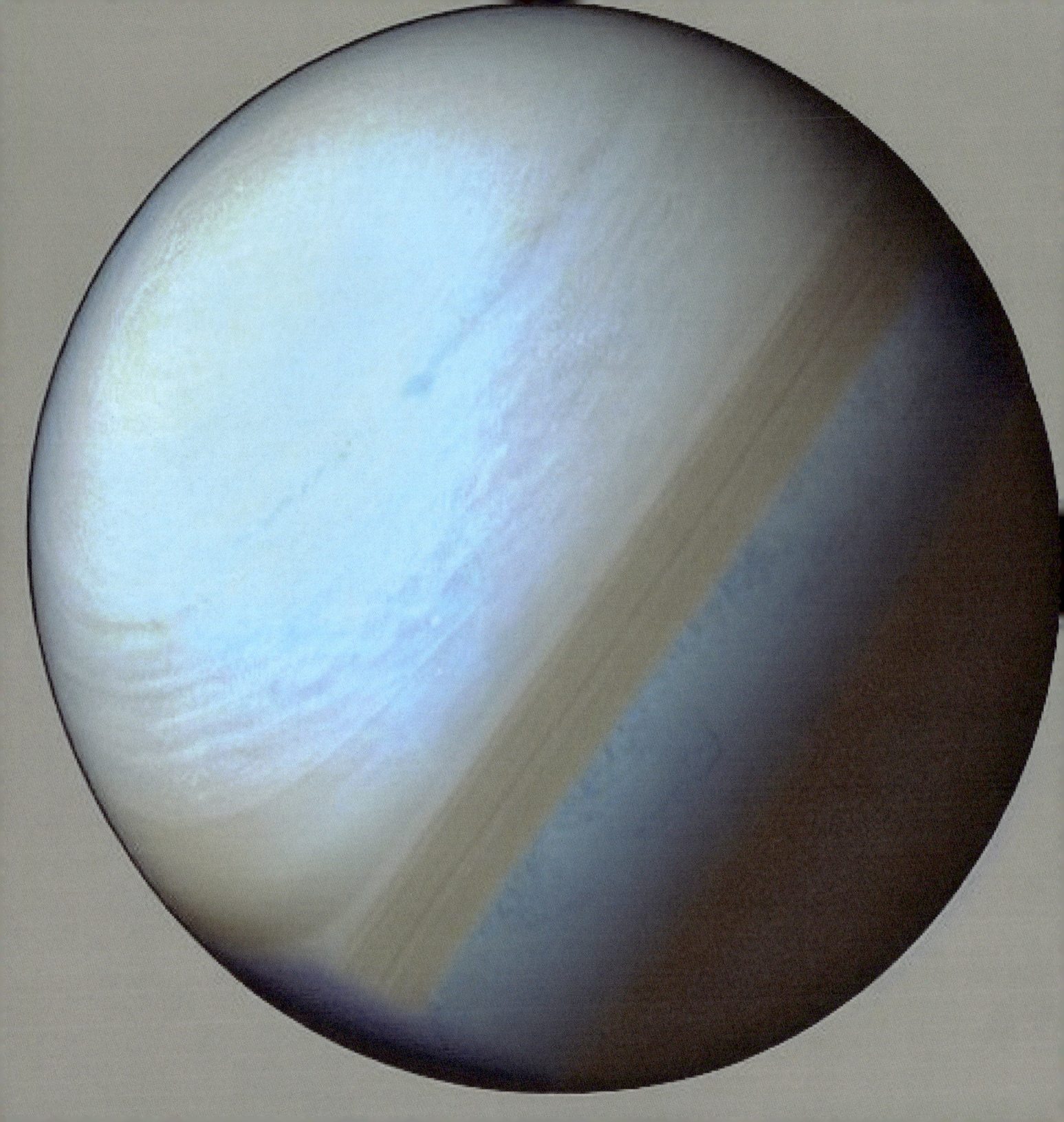

URANUS

Uranus is the seventh planet of the solar system, located after Saturn and before Neptune. It is a gas giant, similar to Jupiter and Saturn, but it has some distinctive features. It is known for its tilted rotation axis, which is inclined by about 98 degrees relative to its orbital plane, meaning that its north pole is almost directly facing the Sun during part of its orbit.

The most noticeable feature of Uranus is its blue-green color, which is the result of the presence of atmospheric gases such as methane, which absorbs red light and reflects blue-green light. Its atmosphere is primarily composed of hydrogen and helium, with traces of methane, ammonia, and water.

Uranus has an internal structure similar to that of Jupiter and Saturn, with a rocky core surrounded by a layer of liquid metallic hydrogen and then an outer layer of gaseous molecular hydrogen. However, unlike Jupiter and Saturn, Uranus does not have a distinct series of atmospheric bands or a Great Red Spot.

Uranus has a ring system, although they are much weaker and less visible than Saturn's. It also has a large family of moons, with 27 confirmed moons, including Titania, Oberon, Umbriel, Ariel, and Miranda. These moons exhibit a variety of geological features, including craters, valleys, and mountains.

Uranus has been visited only once by a spacecraft, the Voyager 2 probe in 1986, which provided the first close-up images of the planet and its moons. However, its study remains challenging due to its great distance from Earth and the limited resources available for interplanetary space missions.

NEPTUNE

Neptune is the outermost of the gas giants in the solar system, located after Uranus and beyond the orbit of Saturn. It is the fourth planet in terms of size and mass. Neptune is characterized by a bluish-green coloration, due to the presence of methane in the atmosphere, which absorbs the red wavelengths of sunlight and reflects the blue ones.

Neptune's atmosphere is primarily composed of hydrogen and helium, with a significant percentage of methane, which gives the planet its characteristic blue color. Its atmosphere is subject to strong winds, which can reach extreme speeds, over 2,100 kilometers per hour, causing dynamic atmospheric features such as spiral clouds and storms.

Neptune has a ring system, although they are much weaker and less visible than Saturn's. It also has a large family of moons, with 14 confirmed moons and several others under study and discovery. Neptune's largest moon is Triton, which is interesting because it orbits the planet in a retrograde direction, suggesting that it may be a captured object.

Neptune's internal structure is similar to that of the other gas giants, with a rocky core surrounded by a layer of liquid metallic hydrogen and then an outer layer of gaseous molecular hydrogen.

Neptune has been visited only once by a spacecraft, Voyager 2, in 1989. The mission provided a valuable collection of data and images that have greatly contributed to our understanding of this remote blue giant and its moon system.

ASTEROID BEAM

In the Solar System, there is a region called the asteroid belt, located between the orbits of Mars and Jupiter. It is a vast region of the solar system populated by numerous small rocky bodies. This belt is characterized by a great variety of sizes, shapes, and compositions of the asteroids that compose it. Asteroids can range in size from small rocks to objects of considerable size, such as the large asteroid Ceres, which is also considered a dwarf planet.

The asteroids in the asteroid belt are mainly rocks and metals, with some variations in composition that may include organic materials and ice. These celestial bodies are the primordial remnants of the protoplanetary disk that gave rise to the solar system about 4.6 billion years ago and represent an important field of study for scientists interested in the formation and evolution of the solar system.

The asteroid belt contains millions of asteroids, orbiting the Sun in a vast expanse of space. Despite their abundance, asteroids in the belt do not frequently collide with each other due to the enormous distances between them. However, occasional collisions can still occur and have played a significant role in the formation and distribution of asteroids in the belt.

THE SATELLITES OF THE PLANETS OF THE SOLAR SYSTEM

The satellites of the planets in the solar system constitute a fascinating and diverse world of moons, each with its own history and unique characteristics. These celestial companions, large and small, play an important role in our solar system, influencing the orbital dynamics of the planets and providing valuable study opportunities for scientists.

Each planet has its own group of satellites, which can vary in number and size. For example, Jupiter, the largest planet in the solar system, boasts a vast collection of moons, including the famous Galilean satellites and many others smaller ones. In contrast, planets like Venus and Mercury have no natural satellites.

Studying satellites provides valuable information about the formation and evolution of the solar system. The geological and atmospheric characteristics of moons can offer clues about the environmental conditions and history of their parent planets. Additionally, many moons may harbor water or other resources that could be useful for future human space missions.

Exploring the satellites of the solar system is an important part of our quest to better understand our cosmic neighborhood and the possibilities of life beyond Earth. Each new space mission reveals new secrets and brings us closer to answering fundamental questions about the universe that surrounds us.

THE MOON

The Moon, our faithful natural satellite, is a world of scientific wonders. Its surface is characterized by a vast variety of geological formations, ranging from basalt plains to mountain ranges, from craters to highlands. These features reflect a history of asteroid impacts and past volcanism.

The Moon lacks a significant atmosphere, meaning it is directly exposed to space and solar radiation. This has helped preserve surface features over time, without the atmospheric erosion process that occurs on Earth. However, it also means that the Moon lacks the protective filter of the atmosphere against cosmic rays and other harmful space effects.

The chemical composition of the Moon is similar to that of the Earth's crust, with abundance of silicates, aluminum, calcium, magnesium, and iron. However, the Moon is relatively depleted in volatiles such as water and light gases, due to the lack of atmosphere and direct exposure to the vacuum of space.

The Moon orbits around the Earth in a period of about 27.3 days, always showing the same face to us due to its synchronized rotation. This phenomenon has intrigued scientists for centuries and has led to a series of theories about its formation and evolution.

The Moon also plays a crucial role in natural phenomena on Earth, such as ocean tides, influencing our climate and the Earth's environment as a whole.

In summary, the Moon is much more than just a rocky sphere orbiting Earth. It is a natural laboratory, a geological archive, and an endless source of inspiration for scientists and space explorers.

THE OTHER SATELLITES

Mars, the red planet, has two small satellites: Phobos and Deimos. These two bodies are so small that their shape has been influenced by tidal forces, making them irregular and asymmetric. Phobos, the larger of the two, is so close to Mars that it appears larger than the Sun in the Martian sky, while Deimos is farther away and appears as a bright star.

Jupiter, the gas giant, is surrounded by a vast system of satellites, including the four large Galilean satellites discovered by Galileo Galilei in 1610: Io, Europa, Ganymede, and Callisto. These satellites are among the largest and most interesting in the solar system. Io is known for its active volcanoes, Europa has a subsurface ocean that could host life, Ganymede is the largest moon in the solar system, while Callisto is covered in impact craters.

Saturn, with its spectacular rings, also has a rich collection of satellites. The largest of these is Titan, an intriguing moon with a dense atmosphere of nitrogen and hydrocarbons. Other Saturnian satellites include Enceladus, with its ice geysers, and Rhea, which has an extremely cratered surface.

Uranus and Neptune, the outer planets, also host a variety of satellites. Miranda, one of Uranus' moons, has a surface characterized by canyons and cliffs, suggesting a tumultuous past. Triton, Neptune's largest moon, is a geologically active world with geysers of liquid nitrogen rising from its icy surface.

This is just a brief glimpse into the diversity of the solar system's satellites. Each moon is a world unto itself, with its own history and peculiarities. Studying these celestial companions provides valuable insights into the origin and evolution of our solar system, as well as suggesting the possibility of habitable worlds beyond Earth.

END OF OUR JOURNEY!

As we come to the end of this thrilling journey through our solar system, it's time to reflect on what we've learned and to celebrate the wonder and complexity of the universe that surrounds us. Through the pages of this book, we've explored the planets, satellites, comets, and asteroids that make up our solar system, discovering their beauty and significance in the cosmic dance that surrounds us.

But our journey doesn't end here. The universe is a vast and mysterious place, rich with wonders yet to be discovered. As we conclude this adventure, let our curiosity and spirit of exploration guide us to new horizons, toward the unknown and the incredible.

May this book have been just the beginning of your journey into the universe, an invitation to continue exploring, learning, and being amazed by the beauty and grandeur of the cosmos. May you always carry with you the wonder and gratitude for our extraordinary solar system and the universe that envelops us. And may you continue to look up at the stars with eyes full of hope and wonder, ready to embark on new adventures in the infinite cosmic space. Safe travels, space explorers!

Printed in Dunstable, United Kingdom